Chinese Kites

Liu Bin

China Pictorial Press · Beijing

Chinese Kites

Liu Bin

First English Edition 2019

By China Pictorial Press Co., Ltd.

CHINA INTERNATIONAL PUBLISHING GROUP

Copyright© China Pictorial Press Co., Ltd.

All rights reserved.

No part of this publication may be reproduced, stored in a retrieval system, or transmitted in any form or by any means, electronic, mechanical, photocopying, recording, or otherwise, without the prior written permission of China Pictorial Press Co., Ltd., except for the inclusion of brief quotations in an acknowledged review.

Address: 33 Chegongzhuang Xilu, Haidian District, Beijing, 100048, China

ISBN 978-7-5146-1791-7

Preface

I'm very happy and honored to write the preface for Mr. Liu Bin's book Chinese Kites. This book expounds on the history and development of Chinese kites from different levels and angles, enabling readers to understand China's special kite culture without limitation in time and space.

With wonderful and extensive pictures and texts, this book fills a gap in the literature on kites, and it will hopefully change some inherent views on Chinese kites. This book will surely change your original cognition of a kite as simply a "flying toy".

As a kite master, Liu Bin has made a comprehensive and unique analysis of Chinese kites based on rich topics, unique designs, and aesthetics combining Chinese and Western arts, and introduced the kite production process in detail with a vivid text and pictures, presenting the amazing kite culture of China to readers. This book elaborates on the concept, aesthetics, types, production techniques and flying skills of Chinese kites, outlining the amazing kite culture of China in an easy-to-understand way.

Kong Lingmin

China's national inheritor of intangible cultural heritage Cao kite

Author's preface
I and Kites

My relationship with kites is seemingly destined. Since birth, I have always been surrounded by kites. It can be said that kites are closely related to my growth, life and career development.

Four generations of my family have engaged in the research and production of kites.

The first generation, my great-grandfather Liu Changfa (1876–1938) was a craftsman in the Workshop of the Imperial Household, mainly responsible for producing lanterns, fans, kites, etc. He was particularly fond of kites, inheriting the craftsmanship of making court kites. Due to social unrest in the final days of the Qing Dynasty, no one played with kites any longer, and he had to make a living in Tongxian County by selling steamed buns. Every Spring, however, he made some kites to help out with the family expenses. However, many kites and materials were lost in those turbulent days full of internal strife and foreign aggression.

The second generation, my grandfather Liu Huiren (1915–2006) began to learn the court kite production skills at the age of 10. Due to difficulty of making a living by selling kites, he engaged in the work unrelated to kite production until retirement. Although busy working, he never stopped exploring and researching the kite art. Regrettably, many drawings and materials were destroyed during the Cultural Revolution.

The third generation, my father Liu Xuejun (1948–), the sixth child of my grandfather, was influenced by him from an early age. Due to his ingenuity, enthusiasm and strong comprehension, he soon grasped kite craftsmanship, and made many representative kites in cooperation with my grandfather.

The fourth generation is myself and I was born in 1977, having extraordinary comprehension and intelligence in kite design and production due to the influence of the previous generations. During childhood, I was often fascinated at the process of kite making demonstrated by my grandfather and my father. In order to better inherit the skills, I began to learn Chinese traditional arts represented by Chinese paintings at the age of 13 and also systematically studied Western art. I successively graduated from Xu Beihong High School of Fine Arts and Xu Beihong Arts College, and finally Renmin University of China, majoring in graphic art.

2003 is a special year.

In 2003, my grandfather, my father and I jointly created the kite brand "Sanshizhai", hoping to inherit and develop the old kite craftsmanship in an organized way. We hope our family brand can be known, recognized and loved by more and more people in China and the world, so as to better inherit the ancient and exquisite kite craftsmanship and culture.

In order to promote my kite-making skills, I learned from several kite masters, including Mr. Tang Jinkun, who, as a bosom friend of my grandfather and an inheritor of the temple-style Cao kite in Tianjin, enjoys a high reputation in the field of kite framework production. In 2006, my grandfather recommended me to Mr. Tang, who helped me achieve significant progress in making of kite framework and special mortise and tenon joint structure, and in perception.

In 2010, I was honored to become an apprentice of the well-known kite master Kong Lingmin, whose works are collected by the National Art Museum of China. In the past few years, he has meticulously and systematically taught me the skills to make swallow-tailed kites and kite patterns, greatly elevating my kite craftsmanship.

With my great love for kites, I have visited many famous professors, intangible cultural heritage inheritors and talented young people in the fields of fine arts, design and education, etc. I have benefited a lot in the process of exchanging ideas with them about design, career and life.

I've also gained knowledge of painting, calligraphy, design, photography, sculpture, porcelain, gold and silver ware, lacquerware, engraving, *Tai Chi*, etc., which has helped promote my design, production and innovation of kites.

I have been committed to achieving the perfect combination of traditional kite craftsmanship with modern technology and techniques. While learning and mastering traditional skills, we have introduced some advanced computer design concepts and equipment, such as hand-painted LCD digital screens, 3D printers, 3D scanners and other digital tools. Bidding farewell to traditional manual design, we use modern computer art and various arts to re-integrate the structure, pattern and production process of kite, increasing the cultural and artistic value of kite, and changing kite toy into kite art.

Appreciate, make, and play.

Learn, innovate, and carry forward.

Thus, it can be seen that te kite has been deeply rooted in my life and daily life.

Contents

Preface 3

Author's preface: I and Kites 5

Chapter 1   Overview

    1. Chinese kite types   001
    2. Aesthetic appreciation   008
    3. Regional characteristics   014

Chapter 2   Brief History

    1. Germination stage   019
    2. Development stage   019
    3. Golden era   021
    4. Modern times   022

Chapter 3   Tools and Materials

    1. Framework   029
        Framework materials   029
        Cutting tools   032

Heating tools   034

Auxiliary tools   034

2. Painting surface   037

Painting surface materials   037

Painting tools   039

3. Pasting and line tying   042
4. Flying   043

Lines   043

Reels   045

Chapter 4   Basic Making Techniques

1. Making the framework   048

Prepare the material   049

Making the components   054

Assembly components   058

2. Drawing the painting surface   063
3. Pasting   066
4. Kuihuo (armor process)   068
5. Tying lines   071
6. Understanding the framework structure diagram   075

Chapter 5   Kite Cases

1. Phoenix-shaped kite   078
2. Goldfish-shaped kite   082
3. Kite shaped like the Monkey King's facial makeup   087
4. Eight trigrams-shaped kite   092

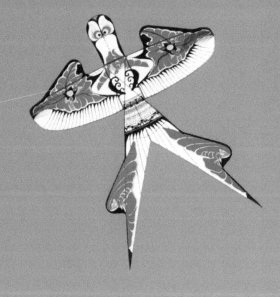

Chapter 1
Overview

The Chinese kite is not only a beautiful flyer, but also an outstanding craft.

A perfect Chinese kite has a delicate framework, a beautiful painting surface, and also an ability to fly. The painted surface composed of lines and colors is the main aesthetic object.

Chinese kites vary in shape, size, geography, and artistic features. This book introduces Chinese kites with northern and southern styles respectively.

1. Chinese kite types

By framework, Chinese kites are divided into the following four types.

Board kites: This refer to plane kites without wing structure, and are divided into soft board kites (Figures 1 and 2) and hard board kites (Figure 3) by edge. The edges of a hard board kite are made up of hard bamboo canes, and board kites can freely express a variety of themes.

Soft-winged kites: This consists of a central body and a pair (multiple pairs) of wing-shaped structures. There is a hard bamboo frame on the upper side (but not the lower side) of the winged structure, which is why it is called a "soft-winged kite". The kite with a pair of wing-shaped structures is called "soft single-winged kite" (Figure 4), and the kite with multiple pairs of wing-shaped structures is called a "soft complex-winged kite". Limited by the contours, soft-winged kites are mainly used to portray birds and insects (Figure 5).

Hard-winged kites: This generally consists of a central body and a pair (multiple pairs) of

Figure 1

Figure 2

Figure 3

Figure 4

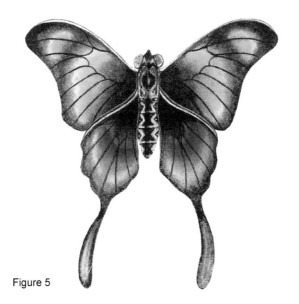

Figure 5

wing-shaped structures. Hard bamboo frames form the upper and lower sides of winged structure, hence its name. Hard-winged kites are divided into different kinds by shape, such as rice-shaped (米) hard-winged kites, glyph-shaped hard-winged kites and bank swallow-shaped hard-winged kites, with different patterns.

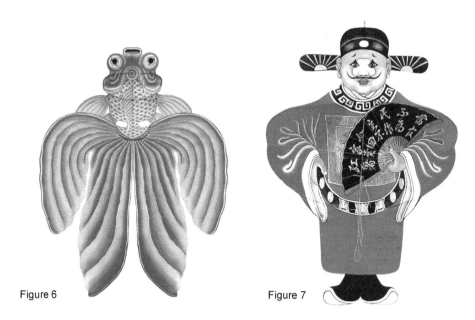

Figure 6

Figure 7

Figure 8

Figure 9

Figure 10

　　A rice-shaped hard-winged kite is generally single-winged, the wing line and bamboo framework of the central body form the shape of the Chinese character "米" (Figures 6 and 7).

　　Bank swallow-shaped kites are the representative hard-winged kites in China. The head of a bank swallow-shaped kite is square, but the top of the head is circular arc-shaped; two wings, chest and abdomen are covered by the hard wing of fixed shape; the tail is fork-shaped and composed of bamboo canes (Figures 8 and 9).

　　Glyph-shaped hard-winged kites are shaped like Chinese characters, and are mostly red. They are symmetrical or asymmetrical. It's extremely difficult to make asymmetrical glyph-shaped hard-winged kites (Figure 10).

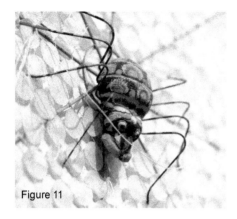

Figure 11

Sieve-shaped kites: These are special, and similar to hard board kites in overall structure. The painted surface is not a whole, but is composed of many independent single pieces connected to the framework by glue and lines. Some three-dimensional paper attachments are usually pasted as decorations (Figures 11 and 12).

Figure 12

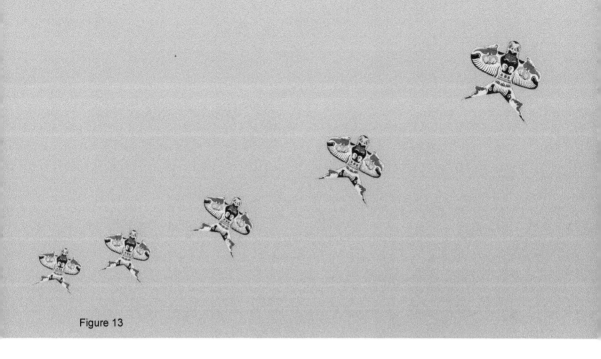

Figure 13

Long-string kites: This comprises a series of identical or similar kite pieces. Long-string kites are usually divided into dragon-string kites and bank-swallow string kites.

A bank-swallow string kite is relatively simple, and is composed of many bank swallow-shaped sections (Figure 13).

A dragon-string kite consists of two parts, namely dragon head and dragon body. The former mainly plays a decorative role, and cannot fly by itself. The latter is composed of multiple identical segments, each of which can fly separately or collectively (Figure 14).

Figure 14

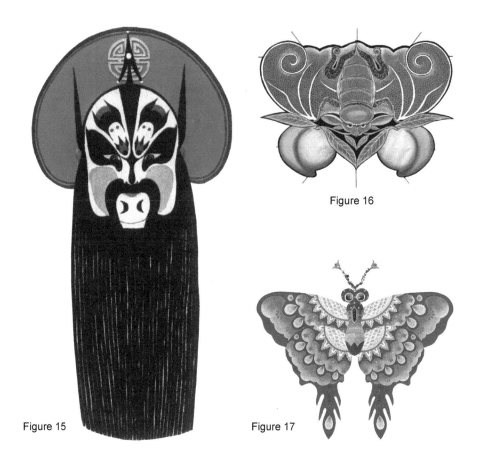

Figure 15　　　Figure 16　　　Figure 17

2. Aesthetic appreciation

The painted surface of a kite should look nice both close-up and while soaring high in the sky. Kites are well designed to achieve the best ornamental effect in both static and dynamic states. The painted surface of kite is usually supported by the framework, and all artistic means including pattern design and coloring are needed to present the unique artistic effect.

Painted surface themes: For Chinese kites, these are very rich and symbolic, usually covering collectables (Figure 15), birds, auspicious patterns (Figure 16), insects (Figure 17), mythological figures (Figure 18), fish and crabs (Figure 19), classical architecture, and so on.

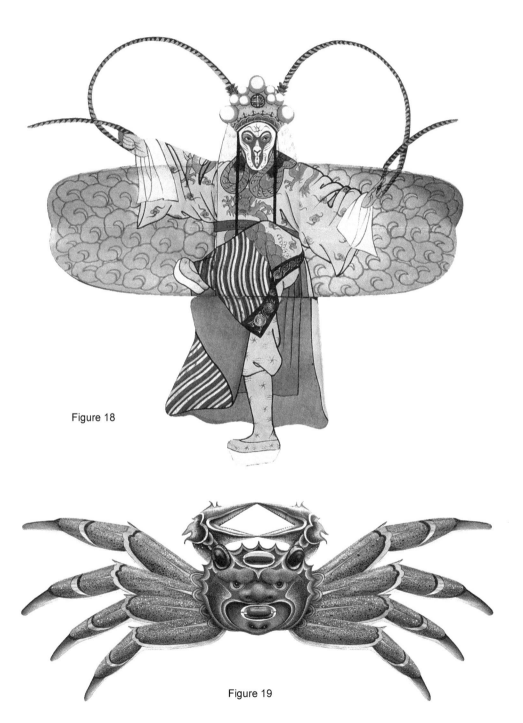

Figure 18

Figure 19

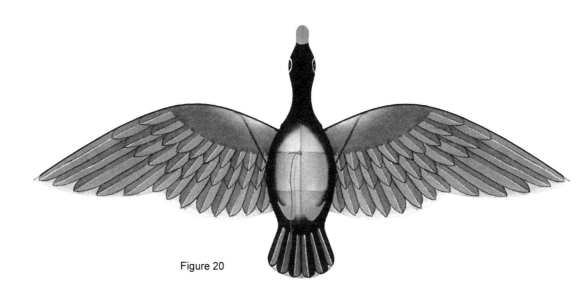

Figure 20

Painted surface shapes: When a kite flies high, the painted surface looks blurry in details, except for its contour. The contour always serves the theme, so different kites might have different contours. For example, the kites with bird and insect themes are generally shaped like their counterparts in nature. No matter how high kites fly, they can be identified at a glance according to their shapes such as bank swallow, swan (Figure 20), and goldfish (Figure 21). Chinese kites are very rich and variable in shape, which is unique in the world.

Painted surface patterns: After the theme and shape are determined, patterns will be

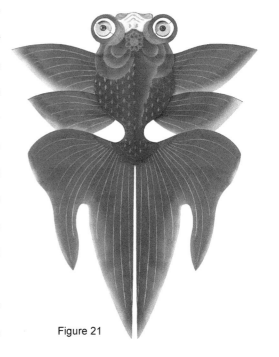

Figure 21

drawn according to the features. Chinese kites have a unique style in composition and color, i.e., generally exaggerating on the basis of realism

Sometimes some decorative patterns derived from the traditional auspicious culture of China are used to express people's good hopes and wishes. The techniques of exaggeration and colorful painting enrich the themes and patterns, conveying strong vitality and unique ethnic styles. Sometimes, even if the added patterns are not very relevant to the theme of a kite, they have rich intrinsic meanings and highlight the theme. For example, some "waist lines" are usually arranged between the chest and abdomen and the tail of a sand swallow-shaped kite. For a fish-shaped kite, a circular pattern is often embedded between the gills and the head, to ensure they are connected naturally, and to set off the theme and pattern (Figure 22).

Painted surface colors: These are generally bright and harmonious, and can vividly express the theme from any angle. In summary, colors are selected in the following four ways.

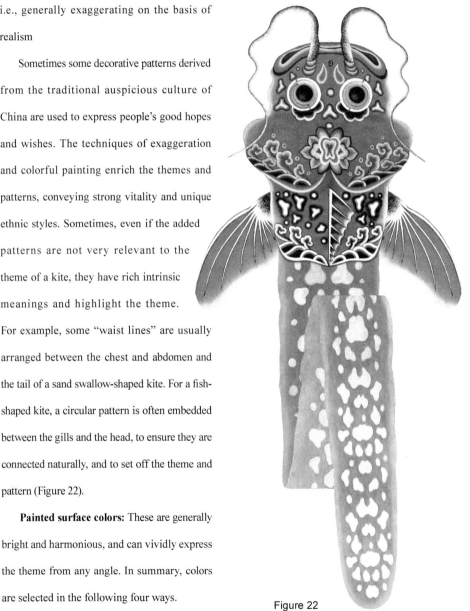

Figure 22

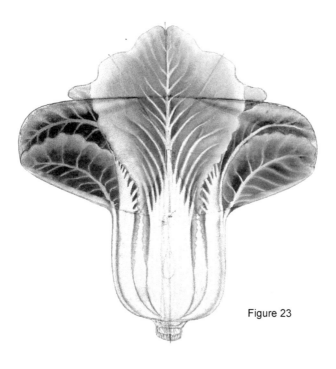

Figure 23

First, a realistic painting approach is optional, that is, colors of material objects (Figure 23) are adopted. For example, the eagle-shaped kite pattern is usually drawn in ochre-brown and black, and the dragonfly-shaped kite pattern is drawn in red, green, yellow or black.

Second, a freehand painting approach is available, i.e., colors different from those of material objects are adopted to create certain meaning or artistic conception. For example, the surface color of a bat-shaped kite is usually vermilion, because "red bat" (红蝠) is homophonic to great happiness (洪福).

Third, visual effect is enhanced through color contrast. Because the surface is only black and white, a kind of bank swallow-shaped kites is commonly known as a "black pot bottom". Such a kite flying high in the sky will have a very strong color contrast. Used to express grief for the death of royal family members in the beginning, it became popular because of its unique beauty. There is also a kind of "blue pot bottom" kite in pure blue.

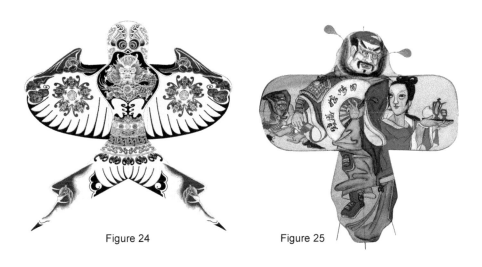

Figure 24

Figure 25

Fourth, the visual effect of flying is enhanced through exaggeration of the color area size. Take the "black pot bottom" for example: Although the exaggerated white claw part is much smaller than the black chest part, it looks very obvious when the kite flies high (Figure 24). The red pattern on the Zhong Kui-shaped kite (Figure 25), and green frog pattern on the chest and abdomen part of Liuheyan kite (Figure 26), etc., highlight the kite themes under the blue sky.

Figure 26

Chapter 1　Overview　013

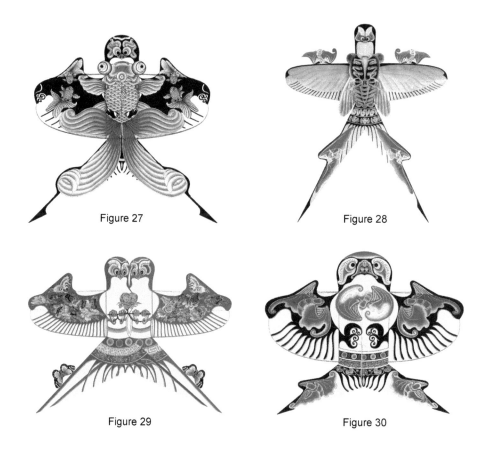

Figure 27

Figure 28

Figure 29

Figure 30

3. Regional characteristics

Chinese kites are rich in shape, variety and style, showing different artistic characteristics due to differences in region and customs. Here, different characteristics of kites from China's four major kite producing areas (Beijing, Tianjin, Weifang of Shandong Province, Nantong of Jiangsu Province) will be introduced.

Beijing-style kites: These highlight the unity of form and spirit (Figures 27-30), while shape and pattern both serve the theme. The processes cover making, pasting, painting and flying. Basic shapes of Beijing-style kites mainly include board, soft wing, hard wing and long string, and patterns are harmoniously integrated with shapes. The sand swallow-shaped kite as a hard-winged kite is a typical representative showing the highest level and creative concept of the Beijing-style kites.

Tianjin-style kites: These have a unique style, absorbing features of Yangliuqing New Year Pictures, Chinese painting and printmaking, and have thick, bright colors and simple lines. The framework is fixed with a tenon rather than line, so that Tianjin-style kites look exquisite, flexible and graceful. The patterns on the painted surface include figures (Figure 31), flowers and birds (Figure 32), insects and fish, etc. Tianjin-style kites are divided into rice-shaped hard-winged kites and soft-winged kites. In particular, the unique soft-winged portfolio kite of Tianjin is composed of multiple soft-winged kite components, each of which cannot fly alone (Figure 33).

Weifang-style kites: Weifang-style kites are exquisite with an appealing

Figure 31

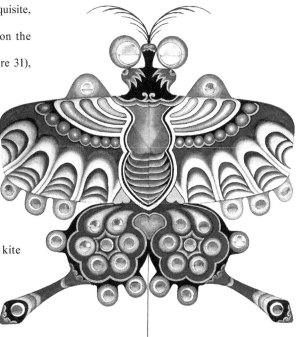

Figure 32

Figure 33

design, and are similar to New Year wood-block painting in painting style. Sometimes, woodblock-printed monochromatic paper is directly pasted and colored. With bright colors, Weifang-style kites are divided into soft-winged kites and rice-shaped hard-winged kites (Figure 34).

Nantong-style kites: Nantong-style kites are exquisite in craftsmanship and shape, and unique in sound, and are mainly represented by "Banyao". As a hard-board kite, Banyao is generally composed of rectangle, square, and rhombus, with exposed hexagon, so it is commonly known as a "hexagonal sparrow hawk" (Figure 35). The length/width of a Banyao kite ranges from less than 1 meters to about 7 meters. With different sizes of "whistle" made of bamboo, wood, and husk, a Banyao kite can make a sound when it flies. The patterns on the "whistle" are different.

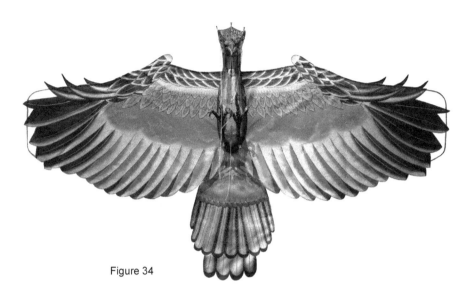
Figure 34

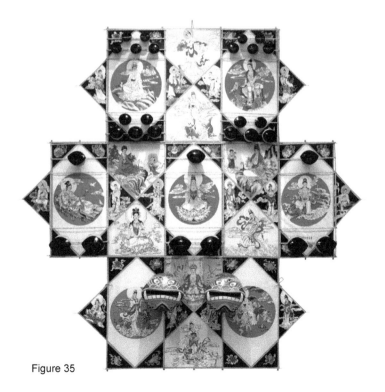
Figure 35

Chapter 2
Brief History

A kite's framework made of slender bamboo and wood, is pasted with thin silk or paper. Kites are easily broken in the process of use, and will decay as time passes. Therefore, there are almost no cultural relics of kites, and most historical materials are derived from text descriptions and paintings.

1. Germination stage

The most primitive kites are toy-like flyers invented by skilled craftsmen.

According to historical records, the earliest Chinese kites were the "wooden glede" made by Mo Tzu (ca. 468–376 BCE) and the "magpie" made by Lu Pan (507–440 BCE) during the Spring and Autumn Period (770–476 BCE).

2. Development stage

During the Tang Dynasty, kites were mainly used for entertainment, or, on occasions, for military communications. As recorded in *Xi Deng Yao Wen*, around the year 784, lantern kites emerged in ordinary society. A lantern kite available then comprised a three-dimensional tubular bamboo rib pasted with silk gauze or bright-colored paper, with a lantern inside the bamboo rib, and a kite line. With help of hot gas and the wind, the lantern kite could rise to the sky, emitting flashing candlelight. Powerful families and honorable clans like this kind of novel kite to show off. However, bad weather conditions or improper operation might cause a fire, so that it was soon banned by the court.

In the Song Dynasty (960–1279), the art of painting flourished. Many paintings of the Song Dynasty on kites are of reference significance for understanding the development process, shapes and structures of kites at that time. In his painting *One Hundred Children Playing in the Spring*, Su Hanchen (1094–1172) portrayed scenes of children playing, including the scene of flying kites; in his painting *Itinerant Peddler*, Li Song (1166–1243) of the Southern Song Dynasty, painted a scene of a woman and her children joyfully purchasing things from an itinerant peddler. Inside the pack was a board-like kite, indicating that they were very popular in the Song Dynasty.

From his childhood, Emperor Huizong Zhao Ji (1082–1135) of the Northern Song Dynasty liked to fly kites. He even wrote a book *Kites During the Reign of Emperor Xuanhe*, which described the process of kite making and painting, which played a positive role in popularizing kites at that time.

In the late 12th century, kite-fighting matches were held. The famous litterateur Zhou Mi (1232–ca. 1298) of the Southern Song Dynasty wrote a book *Wulin Jiushi* describing Lin'an, capital of the Southern Song Dynasty, in which he stated: "Several teenagers on the bridge flew kites, and those whose kite strings were cut off would fail."

At an archeological site of the Jin Dynasty (1115–1234) in Huaide, Jilin Province, a bronze mirror depicting a kite game was unearthed. Four groups of two players each are painted on the reverse side of the mirror (Figure 34). Only one of the

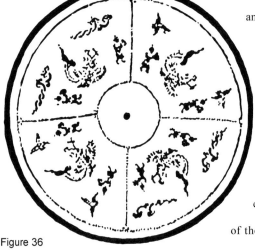

Figure 36

unearthed bronze mirrors had a kite theme, which is also the earliest cultural relic related to ancient kites of China so far.

In the Yuan Dynasty (1279–1368), kites were popular. The poet Xie Zongke (birth and death years unknown) from Nanjing depicted a scene of kite flying: "Each spring, children fly kites on the grassland joyfully. Some kites cannot be taken back because their strings are broken. Looking at the clouds enviously, they hope their kites can float in the sky forever."

3. Golden era

During the Ming Dynasty, the litterateur Xu Wei (1521–1593) described a three-dimensional kite in his poem: "This bird-shaped kite is made up of bamboo canes, and pasted with burlap."

Famous Qing Dynasty writer Cao Xueqin (1715–1763 or 64) wrote the book *Study of Southern and Northern Kites*, which is an important reference material for studying the kites popular in Beijing in the mid-Qing Dynasty, and introduces the formula of kite making and painting in verse, as well as pictures depicting kite colors. In the novel *A Dream of Red Mansions*, Cao Xueqin also described kites many times. For example, in Chapter 70, he introduced eights kinds of kites with the shapes such as butterfly, beauty, phoenix, fish, crab, bat, seven wild geese (connected) and a blood-red "happiness" character, indicating soft-winged kites rose had become very popular at that time. In addition, "seven wild geese" indicates the stringed kites emerged not later than 1760. Another version of *A Dream of Red Mansions* records bank swallow-shaped kites, implying that hard-winged kites had come into existence no later than 1792.

In 1793, Cao Xueqin's contemporary, traditional opera writer Li Dou (?–1817) wrote the book *Notes on a Yangzhou Pleasure-boat*, describing social, economic, and cultural aspects of Yangzhou: "Kites flourished in the Qing and Ming dynasties. Emitting sounds through strings, and keeping a force balance through the tail part, some kites are even 3.33 meters wide and 10 meters long to the tip of their tail. Usually, rectangular, kites are shaped like crab, centipede, dragonfly, butterfly, or the Chinese character for 'fortune', and 'longevity'. Three or five lanterns are tied to the tail part of kites." It can be seen that, in the early Qing Dynasty, board kites still played a dominant role, and simulated and pictographic (besides Banyao) kites were also common.

In 1897, Mr. Jin Junshao wrote the book *Old Spectrum* (Jiu Pu), the earliest monograph of China on kite theory, focusing on the techniques of kite making, pasting, painting and flying, and divided kites into board, soft-winged, hard-winged, Shuangyan, centipede-shaped and miscellaneous kites. This book also mentioned some collapsible kites, reflecting a major breakthrough in the production process. Such collapsible kites are easy to carry and stored, which facilitated the spread of various kites and associated painting skills.

In the Qing Dynasty, the kite trade was recorded in detail. The local chronicle *Dijing Suishi Jisheng* introducing the local customs of Beijing, was published in 1758, which described the kite market in Beijing: "Beijing-style kites are extremely exquisite, and some are very valuable. Kites are traded on the Glass Street." Fang Shuo introduced the situation during the reign of Emperor Daoguang (1821–1850) as follows: "Most of the streets in Beijing are quiet during the Spring Festival. However, an area about one kilometer Glass Street is always bustling…clay figurines, potted landscape, kites, lanterns, and other goods are sold. It is extremely prosperous and always crowded." This record indirectly depicts the influence of kites at that time.

In the Qing Dynasty, Glass Street attracted hordes of tourists from January 1 to January 15 of the lunar calendar. A kite-making industry gradually took shape. At that time, the spacious Glass Street naturally became an ideal place for people to fly kites. A poet once depicted the scene of adults and children flying their kites during the Lantern Festival.

4. Modern times

In 1934, the book *Kites* written by Chen Zefeng was published by Nanjing Zhengzhong

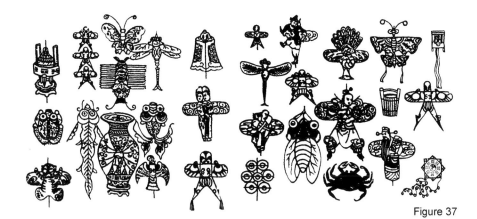

Figure 37

Publishing House, which first introduced the structural features of Chinese traditional hard-winged kites. In the fourth section "Kite Species and General Making Techniques", the author expounded the structures of kites of shapes such as eagle, butterfly, goldfish, and hard-winged hollow characters (such as "土" and "王") (Figure 37).

The development of modern photography techniques made image data of kites available. Some photos record the situation of kite trading in Beijing during the period of the Republic of China. At that time, Beijing-style kites were mainly of the bank swallow-shaped and hard-winged types (Figure 38).

Figure 38

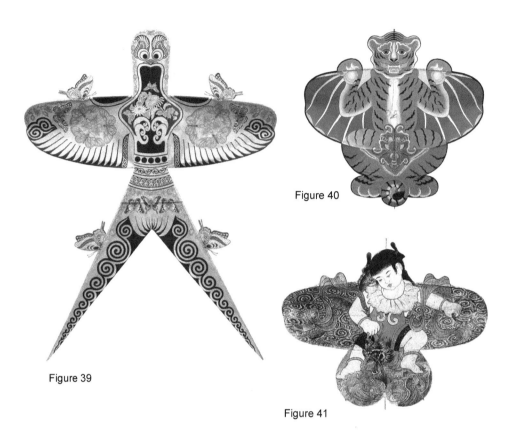

Figure 39

Figure 40

Figure 41

During this era, kite schools were gradually formed in the four major kite producing areas, namely Beijing (Figures 39-41), Tianjin, Weifang, and Nantong.

Two famous kite schools, represented by Ha Guoliang and Kong Xiangze, came into being in Beijing. Among them, Ha Kites are well-known in the capital. So far, four generations of the Ha family have been engaged in the business. The founder, Ha Guoliang, designed the specifications, models and shaping proportions of Ha Kites. At the 1915 Panama Pacific International Exposition in San Francisco, Ha Changying, the inheritor of the tradition, won silver medals and other awards by virtue of the kites shaped like a butterfly, dragonfly, crane, etc. Ha Kites feature rigorous structure, standard size, simple shape and fine patterns.

Kong Xiangze is the founder of Cao Kites. In 1943, he borrowed a manuscript entitled "Collected Manuscripts of the Art-Abandoning Study" from Kaneda, a friend of Takami Kazu who was his Japanese sculpture teacher, and copied part of the book *Study of Southern and Northern Kites*. He also followed the extant kite masters Zhao Yushan, Guan Guangzhi and Jin Zhongnian to learn the techniques. Later, in cooperation with the court kite maker Jin Fuzhong, he reproduced the kites depicted by Cao Xueqin, and called them "Cao Kites". These pursue the complementarity of pattern and shape, unity of form and spirit, and dynamic beauty.

Weifang, Shandong province, is well-known for its handicrafts. Weifang-style kites are unique, combining the characteristics of traditional Chinese painting and New Year wood-block painting. Before the 1980s, Weifang-style kites were not systematic and properly recorded. In 1984, Weifang City successfully held an International Kite Festival. In 1988, Weifang was rated as the "Hometown of Kites", marking that Weifang-style kites had gone global and reached peak popularity. They are mainly shaped like a centipede, various figures, etc. (Figure 42).

Tianjin-style kites are also very distinctive, and representative figures include Wei Yuantai, who was born during the reign of Emperor Tongzhi (1862–1874), and founded a kite making workshop in Tianjin. At the 1915 Panama Pacific International Exposition, he won the gold medal with 11 of his exquisite kites. Tianjin-style kites are characterized by vivid shapes, elegant colors and fine workmanship (Figures 43-44).

Represented by Banyao kites, Nantong-style kites have a long history, as was introduced by Cao Xueqin in the book *Study of Southern and Northern Kites*. Before 1978, Nantong-style kites were in a primary development stage, and were not systematic and properly recorded. Afterwards, some artists and scholars began to systematically record and study their development, publishing a series of books related to kites. With features such as harmonious color, pleasing sound and balance of strength, Nantong-style kites have gradually been exported to other countries (Figure 45).

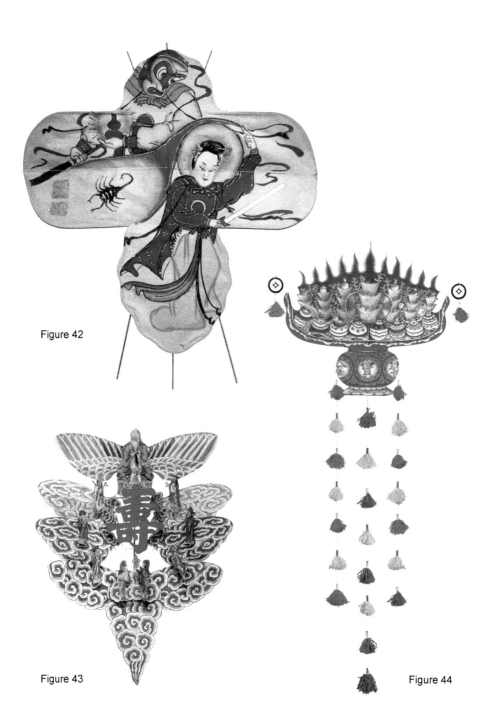

Figure 42

Figure 43

Figure 44

Figure 45

Chapter 3
Tools and Materials

As the Chinese saying goes, "When a workman wishes to get his work done well, he must first ensure his tools are sharpened." Quality tools and materials are very important in making a good kite. This chapter will elaborate on kites from the aspects of framework, painting surface, pasting and line tying, and flying.

The performance of a kite in flight mainly depends on its framework. This is generally made of bamboo because it is light, having a certain intensity, elasticity and toughness, so that it can be bent into the desired shape after heating, and will not deform after cooling. The painting surface of a kite reflects its artistic level, and the tools and techniques used are basically the same as those of Chinese painting. Special tools and materials are required for pasting and tying of the lines to integrate the framework, painting surface and bobbin. Finally, we will introduce the tools needed to fly a kite.

1. Framework

A quality framework can be made through use of handy tools to process the right bamboo properly stored for some time.

Framework materials

By cross-section, bamboo materials are divided into bamboo skin and bamboo meat, and bamboo meat can be further divided into bamboo green and bamboo yellow. Bamboo green

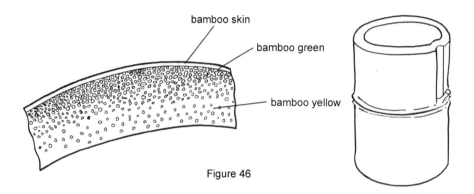

Figure 46

refers to the inner layer of the bamboo meat, having good elasticity; bamboo yellow refers to the outer layer of the bamboo meat, and is hard and crisp, so it is usually removed to make the better framework. Thus, only bamboo skin and bamboo green are left (Figure 46).

Bamboo material with good elasticity and toughness must be chosen. Materials of Chinese kites are mainly moso bamboo and neosinocalamus affinis produced in Fujian, Jiangsu, Zhejiang and Sichuan provinces. In order to facilitate preservation, a fresh bamboo tube about eight meters long is first cut into several segments each about 1.5 meters long, and then is further cut into bamboo clappers about 5 cm wide or bamboo slivers about 3 cm wide, without removal of bamboo yellow. Then, they are put in a dry place not exposed to sunlight for natural drying. Two or three years later, they will turn yellow and can be used (Figure 47). The bamboo newly-chopped off contains a

Figure 47

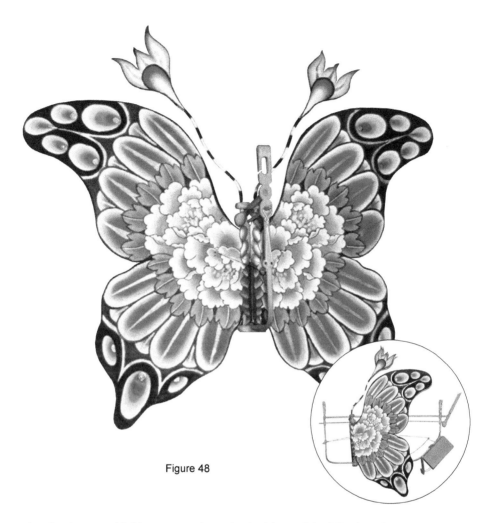

Figure 48

lot of moisture, and is blue-green and poor in elasticity, so it is difficult to shape in the heating process. Even if the framework of a kite is made based on fresh bamboo, it will deform in the flying process due to uneven water volatilization. Of course, using bamboo that is too dry is not suggested because it has poor elasticity and toughness, and is easily broken. To sum up, tender or old bamboo is not ideal for making the framework.

Bamboo material is selected mainly according to the site of a kite to be made, especially the thickness of bamboo yellow and the length of the bamboo section (Figure 48). The bamboo with thick bamboo yellow (5–10 cm) and short bamboo section (20–40 cm), is used to make large kites (more than 1–1.3 meters wide), but it takes a lot of work. The bamboo with thin bamboo yellow (2–3 cm) and long bamboo section (50–80 cm) makes the production of large kites difficult. Use of long bamboo sections helps reduce the workload. Bamboo without a joint can be directly used to make a kite less than 1 meter wide.

Cutting tools

Special knife for bamboo cutting: Used to chop large bamboo clappers.

Single-sided cutter: Used to shovel and peel bamboo.

Double-sided cutter: Used to chop and scrape bamboo.

Electrician's knife: Used to split bamboo.

Hammer: It plays an auxiliary role in splitting thick bamboo. After inserting a knife into the target bamboo, you can hammer the knife back to cut the bamboo more smoothly.

Side-cutting plier: It is used to cut off thin bamboo canes. A side-cutting plier with plain ends must be chosen to ensure a neat fracture.

Bucksaw: It is used to cut wood or bamboo. A large bucksaw and a small one must be prepared. They are used to cut large and thick bamboo and thin bamboo canes respectively.

Wood rasp: Used for rough polishing of bamboo, it is shaped like a flat bar, whose front half part is like a triangular pyramid with coarse granular bulges.

Steel file: Used for fine polishing of bamboo, it is shaped like a stick, whose front is like a rectangular pyramid with fine dimpled grain.

Cutting matt: As a cutting platform, it is placed on the table to protect the blade and the table. Having a certain hardness, it ensures the target can be evenly cut. Although not excessively hard, it can protect the blade. There is no blade slippage when cutting and the surface is non-glare. If it has strong cutting resistance performance, and is used properly, obvious grooves will not appear in one or two years. It should be kept away from any grease, to prevent physical changes that may affect use.

Figure 49

Heating tools

Needle-nose plier: It is used to clamp bamboo and change the bamboo shape in the firing process.

Candle: Emitting small and gentle fire, it is used to shape medium-sized or thinner bamboo canes safely and conveniently.

Alcohol lamp: Emitting a relatively big flame, it is used to bake and shape thick and big bamboo canes. Please use it properly, and do not knock it over or break it, to avoid the risk of fire.

Auxiliary tools

Straight ruler: Needed to make a kite with precise proportions.

Steel tape: Used to measure long objects.

Pointed tweezers: Used to fix the framework of a miniature kite.

Pencil: Used to create engineering drawings, and draw lines and marks on bamboo canes.

Art knife: Used to cut paper or make miniature kites.

Scissors: Used to tailor the painting surface and cut off various lines.

Canvas: Covering your legs when you process bamboo.

Knife grinder: Used to grind tools.

Figure 50

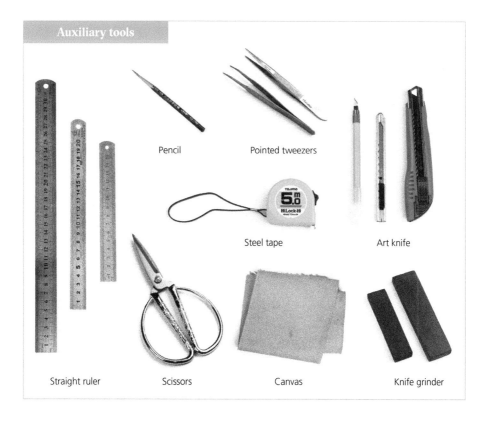

Chapter 3 Tools and Materials

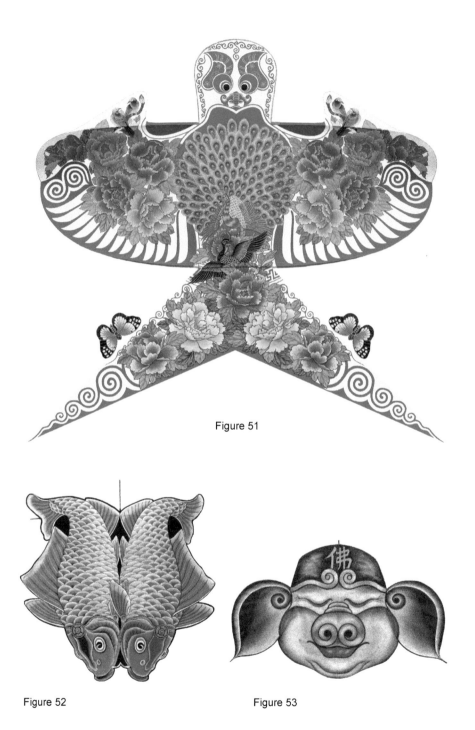

Figure 51

Figure 52

Figure 53

2. Painting surface

Painting surface patterns of Chinese kites are unique and distinctive.

Painting surface materials

Painting surface materials primarily include paper and silk. The following aspects should be considered for selection of appropriate materials:

1) Value: Kites pasted with silk are mostly exquisite crafts or high-grade goods;

2) Nature: Having stronger intensity than paper, silk can withstand a strong wind and can be safely stored for a long time;

3) Coloring and painting: Silk needs to be ironed while painting, which is troublesome. Paper can be ironed after drawing when there is no need of ironing. This makes it convenient and labor-saving;

4) Different materials have different characteristics, which can highlight the kite themes and enhance the ornamental effects. For example, silk with better transparency than paper is used to make the painting surface of a dragonfly-shaped kite, to highlight the transparency of wings.

The material used to make the kite painting surface must be soft but tough, water-absorbing and damage-proof. Technical progress sees the constant emergence of new materials with good performance and low cost. The kites made of such new materials can be preserved for decades.

Xuan paper: Mainly used to make the painting surface of miniature hard-winged kites with a size of less than 40 cm. Despite the rich variety available, few kinds of Xuan paper can be used for making kites. Only that which is soft, tough and not prone to damage, is applicable. It is divided into raw and ripe Xuan paper. The latter is not vulnerable to color bleeding, but can be repeatedly color-saturated and painted. In contrast, raw Xuan has strong performance of water absorption, and the color applied can be scattered soon, so that it is difficult to paint on it. Only those with some experience in use of paper and water are competent to paint on raw Xuan.

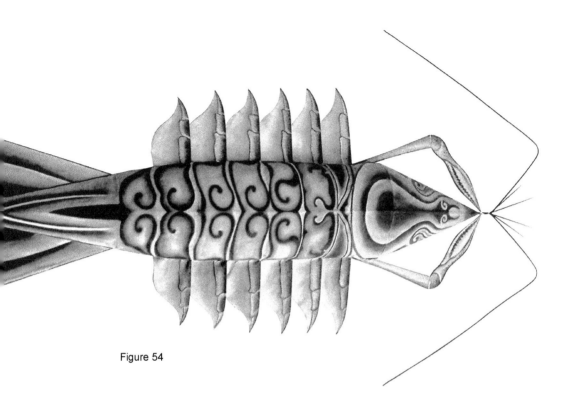

Figure 54

Silk: This was mainly used to make court kites, with exquisite patterns and bright colors. Compared to paper, silk with a long service life makes works more exquisite. Some court kites of the late Qing Dynasty with silk as a constituent element are still preserved in the Palace Museum in Beijing. It should be noted that only this can be used as, like ripe Xuan, is easy to dye, and, secondly, other silk is soft and its ventilation characteristics lead to poor flying performance of kite. Silk is easily applicated in the painting process, and ironing is needed to keep the surface smooth.

Non-woven fabrics: Modern non-woven fabrics of different thickness, can be used to make a painting surface of any kite with the size of less than 1 meter. With good performance in terms of toughness and water absorption, it is soft, cheap and damage-resistant. Despite some color soakage left after painting on non-woven fabrics, they are still more suitable than raw Xuan.

DuPont paper: Modern DuPont Paper is used to make the painting surface of a large hard-winged kite with the size of over 1 meter. It is air-fast, solid, and does not easily decay, and can

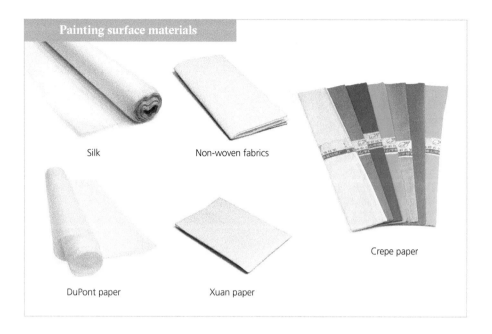

be stored for a long time. As its surface is smooth, acrylic paint is needed for coloring, when bright colors retain their sheen. Although DuPont Paper can be kept level in the painting process, clear marks cannot be left, and ironing is not allowed.

Crepe paper: As a kind of colored pleated paper, it is used to make the decorations maintaining the flying balance of a kite.

Painting tools

Writing brush: These are divided into hard and soft types. The hard brush is used to draw the lines, and the soft brush for coloring. Several different sizes of brushes can be prepared to paint kites of different sizes.

Scrubbing brush: A small or large scrubbing brush is used to draw the painting surface.

Felt: This is a thick wool mat generally white, flat and water-tight. It is used for padding paper or canvas, and can prevent reverse contamination of water-soluble pigments after diffusion on the table.

Figure 55

Palette: Generally made of plastic or ceramics, a palette is divided into a number of separate areas by a raised barrier, and can be used for preparing a plurality of colors.

Brush washer: This container used to clean the brush whose depth should exceed the length of the brush tip. Before changing the color, the brush needs to be cleaned in a large amount of water.

Brush rack: This is a shelf with multiple grooves for placing the brushes temporarily when not in use, which can prevent any pigment on the brushes from contaminating other places.

Paperweight: Mostly made of metal or stone, it is used to press against paper when writing or painting, so as to keep it level.

Pigment: Two kinds of pigments are generally used for the painting surface. First, water-soluble pigment has a very good light transmission effect, which makes the painting surface look light, thin, and colorful. Second, covering pigment (such as gouache and propylene) can

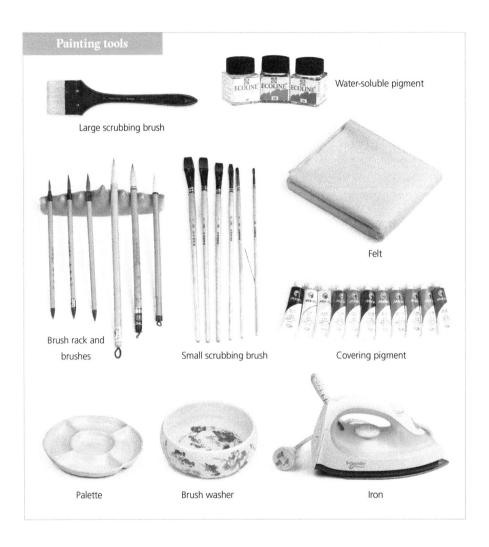

be repeatedly applied to achieve the light-tight effect, but will slightly increase the weight of the kite. The two kinds of pigments should be properly used according to their different characteristics and kite design requirements. In particular, to make the silk-based painting surface lifelike and gorgeous and present a contrast effect of transparency and opacity, two kinds of pigments are generally used for one kite.

Iron: Used to iron the kite fabric to make it level and smooth.

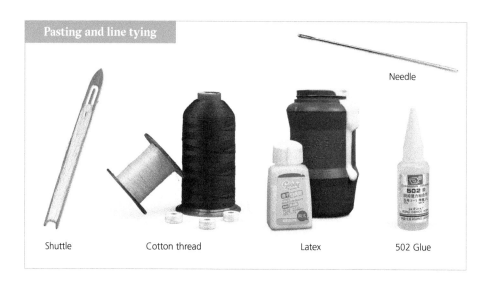

3. Pasting and line tying

Shuttle: It is used to quickly wind a large number of lines when tying a kite.

Cotton thread: This connects the framework to the reel. Do not use plastic fishing lines to prevent possible disconnection.

Needle: Used to help tying lines.

Latex: It can firmly fix the joints between bamboo canes and between them and the kite surface firmly, but takes five hours to solidify.

502 Glue: As a quick-drying adhesive, it can solidify in seconds, and is mainly used to fix the joints between bamboo cane and line. Latex and 502 glue can also be used together to achieve a better effect.

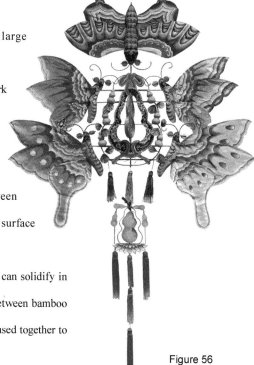

Figure 56

Figure 57

4. Flying

For the kites of different sizes, shapes and structures, different kite lines and reels must be used to ensure a proper flying performance.

Lines

The right line is important for a kite. Generally speaking, the kite line should meet the following requirements:

High strength of extension: The stress on a flying kite often exceed several, 10 or even several dozen times its own weight. Together with air resistance, the stress works on the kite line, so it naturally must have sufficient strength of extension.

Light weight: The lightest line that can withstand the same tensile force is suggested. The materials of lines vary greatly in performance. In addition, other factors such as toughness and wear resistance should be considered.

Toughness and wear resistance: Toughness and wear resistance are required because the kite line is often repeatedly wound in or sometimes rubs against other kite lines or obstacles.

Small air resistance: In general, smooth lines face less air resistance.

Flying tools

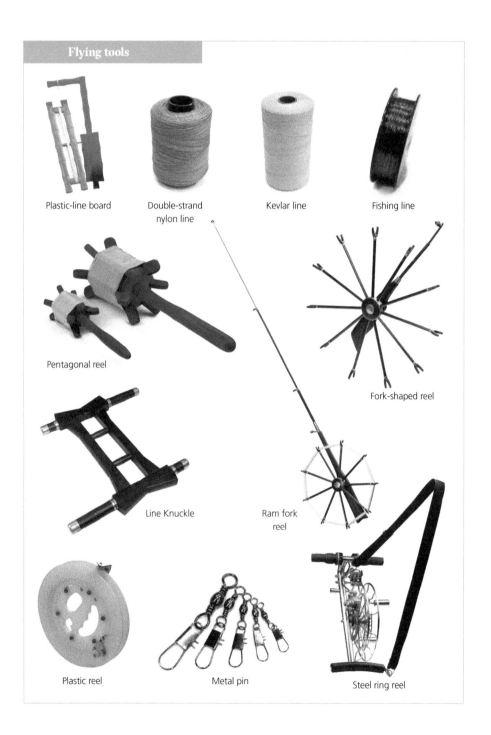

Plastic-line board

Double-strand nylon line

Kevlar line

Fishing line

Pentagonal reel

Fork-shaped reel

Line Knuckle

Ram fork reel

Plastic reel

Metal pin

Steel ring reel

Weak elasticity: Elasticity is divided into tensile and torsional elasticity. The former will affect the timely control of the kite, because of delayed response. The latter is more unfavorable for flying a kite, because the accumulated torsional force beyond the elastic tolerance limit will twist the kite and make it lose balance. Use of torsion joints can avoid the adverse impact of twisting.

The following three types of lines may be chosen according to the kite size:

Double-strand nylon line: Cheap, strong in tensile strength, but not wear-resistant, it is suitable for small and medium-sized kites.

Kevlar line: As a fireproof fiber material, it features strong tensile strength, wear and heat resistance.

Fishing line: Featuring high strength, low resistance, and poor wear resistance, it is too thin, thus easily causing compression damage. If a fishing line is used for a large kite, protective gloves must be worn to prevent hand damage. It is characterized by strong tensile elasticity. It greatly extends at high temperatures, and is brittle and easily broken at low ones.

Reels

The reel should be light, solid, durable, and allow easy withdrawal and release of the lines. Along with technological advances, disc-shaped axles have emerged, sturdier, more durable and easier to operate.

A long-handled hexagonal or pentagonal reel: As a traditional tool, it can withdraw and release lines in a rotating way.

Fork-shaped reel: Suitable for most kites, enabling faster withdrawal and release of lines.

Steel ring reel: Suitable for large kites, with greater strength.

Plastic reel: Suitable for ordinary kites.

Ram fork reel: Suitable for miniature kites.

Plastic-line board: Suitable for simple miniature kites.

Line knuckle: Suitable for super large kites.

Metal pin: Used to connect the reel and the kite, it facilitates assembly and disassembly.

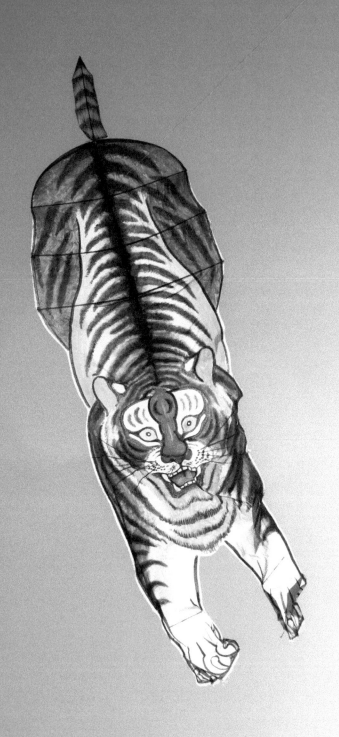

Chapter 4
Basic Making Techniques

Chinese kites go through four stages of manufacture: making the framework, drawing the painting surface, pasting and tying.

First, to make the framework, process bamboo canes with knives, prepare all components needed, combine and fix them to form a framework after firing in an oven.

Second, to draw the painting surface, one should draw the theme pattern on the painting surface material that creates an artistic effect (Figure 58).

Figure 58

Third, pasting to attach the painting surface to the framework.

Fourth, in order to tie lines, a variety of knots are used to connect the kite with the reel.

After understanding the basic techniques, you should learn to understand the framework diagram.

1. Making the framework

The framework is made by three steps: First, prepare the bamboo material for rough processing; second, lay out the bamboo components in line in accordance with the kite design requirements through fine processing; third, assemble the components (bamboo canes), that is, connect them with glue and lines to form the desired framework.

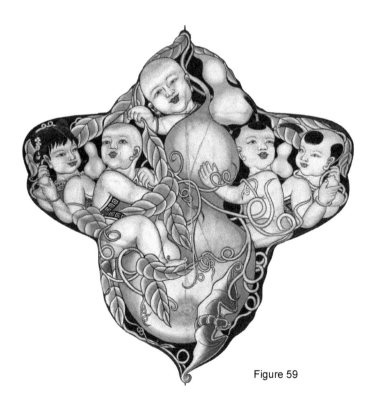

Figure 59

Prepare the material

Before assembling a complete framework, you need to prepare the material through the following carving and firing work: Saw bamboo to the desired length, split for desired thickness, trim and file any bamboo knots, and finally remove the outer skin of bamboo to make it smooth.

Single-sided and double-sided cutters are often used for bamboo processing. According to the difference in posture and precision, processing techniques are divided into splitting, peeling and scraping. After the basic processing, you need to level, straighten and fix the length. If any bamboo cane is curved or uneven, firing is suggested for adjustment. If any bamboo cane is uneven in width and thickness, the excess part should be removed by peeling and scraping. If the length is excessive, you can use a side-cutting plier to pinch off the unwanted part. The finished bamboo canes should be regular in shape, uniform in size and suitable in length.

Cut bamboo to the desired length: Saw and shear

A long bamboo cane can be processed to the desired length by a bucksaw or side-cutting pliers. The bucksaw is used only for larger bamboo canes in the material preparation stage. Side-cutting pliers are generally used to cut bamboo canes with a thickness of less than 1 cm.

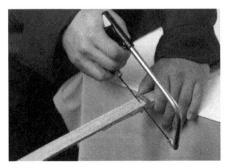

Saw bamboo canes.

Shear bamboo canes.

Process bamboo for the desired thickness: Splitting

"Splitting" refers to the technique of cutting off or cracking a large amount of bamboo vertically. The bamboo near the skin is strong in intensity, so only the part close to the bamboo meat can be removed. If the bamboo thickness is appropriate, the inner skin of bamboo meat should be scraped. In addition to basic splitting, pull-splitting, finger-splitting and prick-splitting can be conducive to achieving different effects.

Basic splitting: A knife is used to vertically split the bamboo cane in the direction consistent with the fiber. If the strength used proves insufficient, you can hammer the knife blade back. The splitting trajectory will extend along the bamboo fiber, and slanting bamboo fiber easily results in deviation. In this case, firing is needed to correct the bamboo fiber. Finally, cut the bamboo to the end to thoroughly split it.

Pull-splitting: If the splitting trajectory tends to deviate, it can be corrected at the initial stage of deviation. When you force apart the bamboo cane on one side backward with a knife, the splitting trajectory will return to the opposite direction. In this way, you can make bamboo canes of different width.

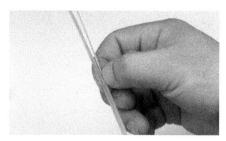

Finger-splitting: This is used to process bamboo cane with straight fiber and the thickness of less than 1 mm, and is very suitable for making miniature kites. First, use a knife to open a crack, and then use your fingernails to split the bamboo along the cracked fiber. Compared to pure use of a single knife, this method is effective for processing very thin bamboo in a faster and more accurate way.

Prick-splitting: First, pierce the middle of the bamboo cane with a knife, maneuver the knife to create an opening so as to change the shape of the bamboo cane. This method is generally used with plug connection (see P70).

Process bamboo for the desired thickness: Peeling

"Peeling" is a technique of obliquely cutting the bamboo cane with a knife mainly to remove its excess part. More refined than "splitting", peeling is divided into embrace-type, push-type, draw-type peeling and handheld peeling.

Embrace-type peeling: Facing the table, put the bamboo cane lengthwise on the cutting matt in front of you, press the bamboo cane with one hand while holding the single-sided cutter with the other, and start to peel inward horizontally while keeping the blade and the bamboo cane at a fixed angle.

Push-type peeling: This is basically similar to embrace-type peeling. The difference lies in that the bamboo cane is peeled forward.

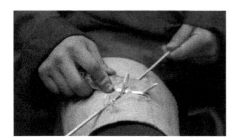

Draw-type peeling: It is suitable for processing long bamboo canes. The process is as follows: Sit down, put a thick canvas on your thigh, put the bamboo cane on the canvas with one hand while holding a single-sided cutter with the other hand; repeatedly pull the bamboo cane while keeping the blade static. There should always be a fixed angle between the blade and the bamboo cane.

Handheld peeling: It is suitable for processing miniature bamboo canes. Grasp the back of the bamboo cane with the hand holding the knife, and then peel until the desired thickness is achieved.

Process bamboo for the desired thickness: Scraping

"Scraping" is more refined than "peeling", functioning to vertically process the bamboo cane. It is generally used to finally remove the outer green skin of the bamboo, so as to reveal the yellow part and make the framework clean and beautiful. The process is divided into horizontal, draw-type and hand-held scraping.

Horizontal scraping: Sit down in front of a table; put the bamboo cane on the cutting mat in front of you; press the bamboo cane with one hand while holding the double-sided cutter with the other hand, and horizontally scrape from the near position while keeping the blade always touching the bamboo cane vertically.

Draw-type scraping: This is more suitable for processing longer bamboo canes. Sit down, put a thick canvas on your thigh, then put the bamboo cane on the canvas with one hand while holding a double-sided cutter with the other hand. Repeatedly pull the bamboo cane while keeping the blade touching the cane vertically.

Hand-held scraping: This is suitable for processing miniature bamboo canes. Grasp the back of the bamboo cane with the hand holding the knife, and then scrape until the bamboo cane reaches the desired thickness.

Figure 60

Baking

Baking is a technique of firing bamboo canes with heat sources such as an alcohol lamp or a candle to achieve shaping. In the firing process, the bamboo canes should be kept away from the flame to a certain distance to prevent ignition or over-toasting. Baking aims to bend or over-fire the canes.

Bending: As the name implies, this aims to bend the bamboo cane through firing to achieve the desired final shape. All sides of the bamboo cane can be bent. Generally, you can hold the sides of the bamboo cane by hand, or use such tools as sharp-nose pliers to prevent scalding. After determining the bending position and the bamboo surface, you need to bake the back of the target surface evenly. After the cane is softened, press downwards and inward with your hands to make it bend to the appropriate angle, and then remove it from the flame to cool down. Repeat this step until a satisfactory shaping effect is achieved. Long heating is not allowed to prevent ignition or over-toasting.

Over-firing: Heating can improve the physical strength of the bamboo cane and fix its shape. Any part of the kite framework should be evenly over-fired and/or bent. All oil and moisture contained in the bamboo cane must be eliminated by baking to prevent deformation of the individual cane and thus the framework.

Trim and file bamboo knots

Bamboo knots are trimmed and filed to make them straight and beautiful, so as to meet the needs in terms of framework shape and structure. Bamboo knots are bent through baking first and then filed to make them straight.

Trimming bamboo knots: **Bend the obverse and reverse sides of the bamboo knot in the firing process, to straighten the bamboo fiber bent around the bamboo knot.**

Filing bamboo knots: **First, use a wood rasp to flatten the convex part of the bamboo skin, and then use a steel file to finely grind it, which is called "filing bamboo knots forward"; turn over the cane, and use the same method to flatten the inner convex part, which is called "filing bamboo knots in reverse". In this way, bamboo knots will become straight.**

Making the components

After preparation of the required bamboo canes, you need to make some special components through fine processing. Some special techniques will be introduced here, including symmetrical splitting, central rod and all-in-one wing strip techniques.

Symmetrical splitting technique

This is used to make two symmetrical framework components that are basically the same. As the most important basic technique for making framework components, it combines splitting, scraping and peeling.

Figure 61

Chapter 4 Basic Making Techniques 055

1. Make a bamboo cane whose width is more than twice that of the required component. According to the design drawing, the whole cane will be baked.

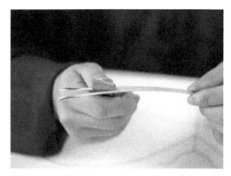

2. Use a single-sided cutter to cut the baked cane in half from the middle.

3. Process one bamboo cane above by peeling and scraping until the width meets the design requirements.

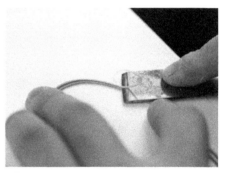

4. then, process the second cane so that it matches the first one as above.

5. Scrape the second cane to make it the same as the first in shape and size.

6. Finish processing the two canes.

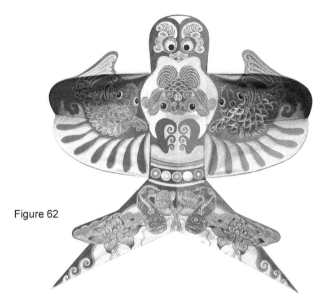

Figure 62

Central rod technique

This is mainly used to make central rods for a variety of kites before baking. It will make the central rod hard in the upper part and soft in the lower part.

The central rod should be made thick (hard) in the upper part and thin (soft) in the lower part by peeling and scraping.

All-in-one wing strip technique

This is used to make a wing strip that is symmetrical on both sides in terms of softness and hardness. It should be made before firing.

The two sides of bamboo should be made consistent in thickness and hardness by peeling and scraping.

Tips for use of the shuttle

The shuttle is generally used to weave the fish net, and facilitates the threading between the kite framework gaps. There is a groove at the lower part of the shuttle, and a tongue-shaped bamboo chip at the center. The shuttle is used as follows: wind the line around the groove at the bottom and the bamboo cane in the middle, and then pull the shuttle with the thread end back and forth.

Assembly components

The prepared bamboo canes should be connected with glue and lines to get the desired framework. Here, we will mainly introduce the methods of connection and binding to assemble framework components independent but closely related.

There are six connective methods: overlap, plug, clamp, butt clamp, kneeling and cohesive connection.

Binding methods are divided into cross and parallel binding. In order to improve efficiency, the shuttle can be used.

Connection methods: Overlap connection, plug connection, clamp connection, butt clamp connection, kneeling connection

Overlap connection: Vertically place two canes across each other.

Plug connection: Vertically insert a bamboo cane into another with an opening in the middle.

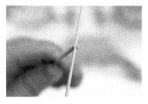

Clamp connection: **Clamp** another bamboo cane vertically with the little opening at one end.

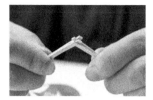

Butt clamp connection: **Clamp** two bamboo canes through their split tails ends.

Kneeling connection: **Connect** one cane with another with one end is baked into a desired arc through the line.

Connection method: Cohesive connection

1 Mark the ends to be connected with a pencil.

2 Cut the cohesive connection ends of two canes to form two inclined surfaces that are the same in shape, opposite in position, and can be matched.

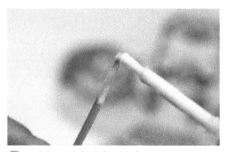

3 Apply latex to the inclined surfaces.

4 Match the two inclined surfaces.

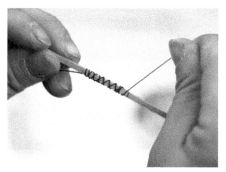

5. Temporarily fix the cohesive connection position with the line, and wait until the glue solidifies.

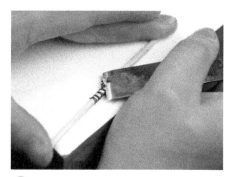

6. After the glue is solid, use the embrace-type peeling method to remove the temporary binding line. Then trim the joints of the two canes.

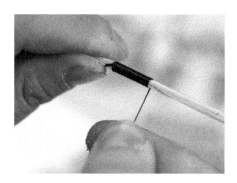

7. Use the parallel binding method to tie the cohesive connection part.

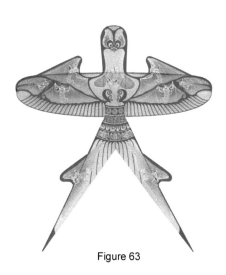

Figure 63

Figure 64

Cross binding

This method is used to cross-bind two bamboo canes, and is suitable for overlap, plug and clamp connections.

1. Press the thread end under the subsequent binding line to fix it in place.

2. Wind the line for the same circles along two diagonal lines.

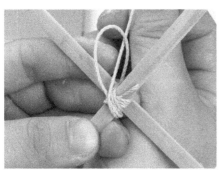

3. Tie a normal knot as the end to be fixed.

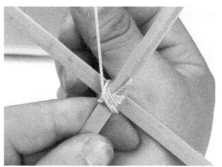

4. After binding, apply a small amount of 502 glue to fix it in place.

Parallel binding

This is used to tie two bamboo canes side by side neatly and tightly, and is applicable to kneeling and cohesive connection.

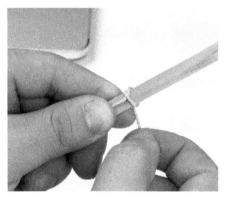

1 Press the thread end under the subsequent binding line to fix the former.

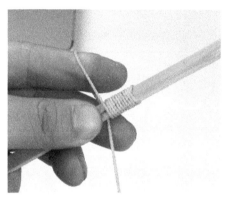

2 Wind the line for a sufficient number of circles.

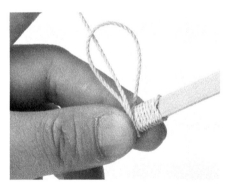

3 Tie a normal knot as the end to fix.

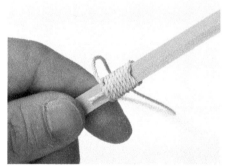

4 After binding, apply a small amount of 502 glue to fix the knot position.

Tips for air drying of the framework

After several years, a bamboo framework may be affected with damp or incomplete evaporation. To completely eradicate this phenomenon, the newly made framework needs to be naturally air-dried for a few days or even months before painting, pasting, etc.

2. Drawing the painting surface

An easy-to-use oil pastel or water color brush, or complex elaborate-style painting techniques can be used to draw the painting surface. Here, only the elaborate-style painting techniques for kites will be briefly introduced.

The painting surface can be drawn through outline drawing and coloring.

Outline drawing

This involves outlining the pattern with black chalk line.

A hard brush with a very thin tip is often used for outline drawing. Most kite patterns are symmetrical from left to right. First, outline half of the pattern on a piece of paper, cover it with another piece of paper after the ink is dried, and draw the other half of the pattern along the previous outline.

Coloring

Coloring is conducted after outline drawing, and is divided into level and saturated types. Level coloring makes the color distribution even, while saturated coloring makes the transition between two different colors show a gradient effect.

After coloring, you need to redraw the outline with the appropriate pigments (usually including, but not limited to, black, white and golden) to present a clear pattern and structure, which can somewhat offset any coloring error. For saturated coloring between a variety of colors, you need to apply different pigments to adjacent color parts with various brushes.

Level coloring: Repeatedly apply the same color of pigment to a certain area to ensure the color is evenly distributed.

Saturated coloring: Use two brushes dipped in pigment and clear water respectively to perform saturated coloring between colored and colorless parts. Paint with the pigment brush first, and then repeatedly apply the water brush before the pigment solidifies to ensure even distribution. A piece of napkin may be used to absorb excess water from the brush. An appropriate amount of water is required for achieving a satisfactory color effect.

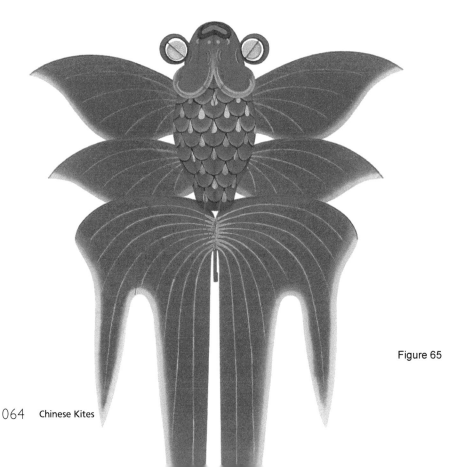

Figure 65

Tips for brush holding

Posture of brush holding varies based on different sizes adopted as necessary.

Basic posture: Press against the brush with the prominent part of your index finger, middle finger and thumb, the brush perpendicular to the paper surface, ensuring it can be flexibly rotated; move the brush with your ring finger and little finger. You should gather together your fingers to firmly fix the brush position, but should leave some space between your palm and fingers to ensure flexible movement of the fingers.

Wrist fixing posture: The brush holding posture is no different from the basic posture. Your wrist and elbow are placed on the table as support so as to limit the scope of the brush movement, but has the advantages of high operational accuracy and low difficulty.

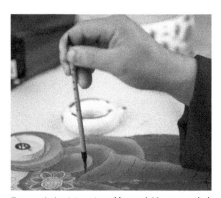

Suspended wrist posture: Your wrist is suspended, while your elbow is placed on the table as a support point, ensuring a larger movable range of brush than that of the wrist fixing posture. However, the operational difficulty will be slightly increased.

Suspended elbow posture: Your wrist and elbow are both suspended, which makes it difficult to achieve the maximum movable range of brush and also increases the operational difficulty. Constant exercise is required to grasp this skill.

3. Pasting

Pasting is followed after preparation of the framework and painting surface. Before pasting a kite, you must prepare the painting surface well according to the framework size. The painting surface should be longer than the outer contour of the framework, and outstretched edges should be proportional to the size of the kite. For example, the outstretched edge of a kite with a horizontal width of 1 meter is usually about 1 cm in length.

The kite body and edges need to be pasted.

Pasting the kite body

Body pasting means adhesion of the back of the painting surface to the outer border of the framework.

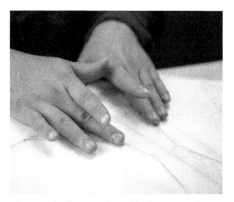

1. Apply latex evenly to the obverse side of the kite framework. The amount should be moderate, and excessive latex will slow down the speed of drying. Then, put the latex-coated framework on the painting surface accurately.

2. Press the framework until he becomes cohesive with the painting surface. While pressing vertically, twist and stretch constantly, and continuously adjust according to the actual situation, to ensure the surface is firm and tight.

Process edges

Edge processing is divided into covering and trimming. Edge covering means wrapping three sides of the bamboo cane with paper. After trimming, part of the bamboo canes will be exposed, which shows the level of bamboo cane production.

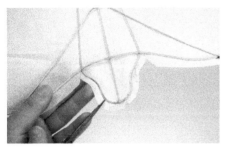

Edge covering step 1: Cut open part of the painting surface exposed on the curved part of the framework.

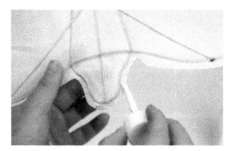

Edge covering step 2: Apply glue on the painting surface.

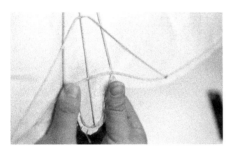

Edge covering step 3: Paste the painting surface onto the framework.

Edge covering step 4: Directly paste the linear framework as shown in the figure.

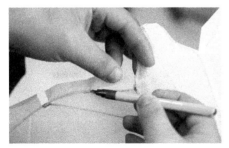

Edge trimming step 1: After the glue becomes solidified, use an art knife to cut off the excess paper on the framework.

Edge trimming step 2: The effect is as shown in the figure.

4. Kuihuo (armor process)

Kuihuo is a special technique of pasting multiple layers of water-soaked paper onto specific positions. When such paper is air-dried and shrunk, tension will increase, which contributes to shaping. The framework is partially wrapped like armor, so it is called "armor process". A new layer of paper cannot be pasted on before the previous layer is at least half-dry; 4–5 layers of paper will be pasted, and be naturally air-dried before the start of next piece of work. Kuihuo is divided into two kinds. See the following texts and pictures for details.

First Kuihuo

Directly paste paper on the framework to shape, and then color.

1 Determine the size of the paper to be used according to the armor area.

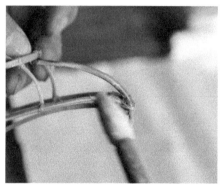

2 Paste the first layer of paper onto the framework in a normal way.

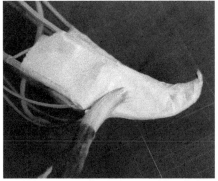

3 Finish pasting the first layer.

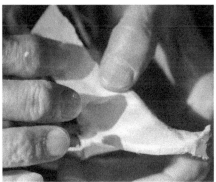

4 When this is half-dry, paste on another layer of paper.

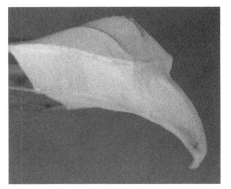

5 Paste 4-5 layers of paper in total.

6 Compared to thin single-layer paper, multiple layers have a better shaping effect and texture.

Chapter 4 Basic Making Techniques

Second Kuihuo

Shape with plaster, then paste paper onto it; color after paper is dry; remove the colored paper and paste it onto the framework. (See "Armor Portrait–1. 2", "Armor Eagle Head–1. 2") The specific technique is the same as that of the first Kuihuo.

Armor Portrait–1. 2

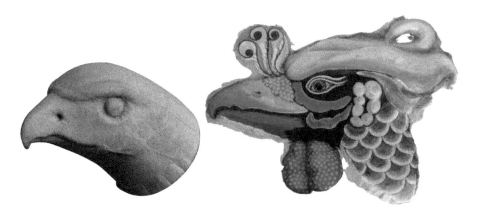

Armor Eagle Head–1. 2

5. Tying lines

The lifting line is used to connect the kite with the reel, which must be firm and balanced to ensure the final safe flying of the kite.

Tying methods of rope knot

Four kinds of rope knots are required: Ordinary, half-folded, bowknot, thief-catching knot.

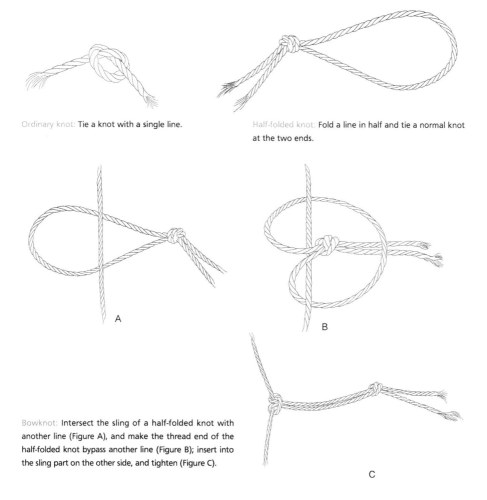

Ordinary knot: Tie a knot with a single line.

Half-folded knot: Fold a line in half and tie a normal knot at the two ends.

Bowknot: Intersect the sling of a half-folded knot with another line (Figure A), and make the thread end of the half-folded knot bypass another line (Figure B); insert into the sling part on the other side, and tighten (Figure C).

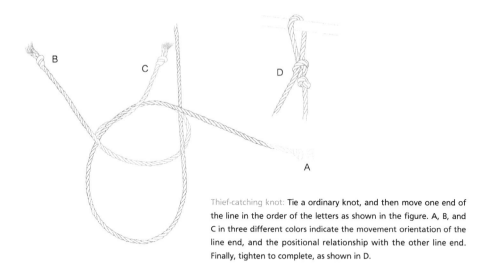

Thief-catching knot: Tie a ordinary knot, and then move one end of the line in the order of the letters as shown in the figure. A, B, and C in three different colors indicate the movement orientation of the line end, and the positional relationship with the other line end. Finally, tighten to complete, as shown in D.

Tying methods

There are three methods for connecting the framework with the reel, which are applicable to different kites.

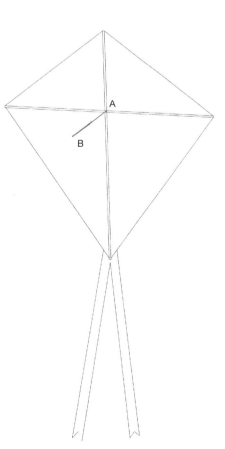

Single-line tying method: As the name implies, a line is used. A thief-catching knot is tied at point A connecting the kite framework, and the half-folded knot is tied at point B. The thread end of the connecting the reel should be prepared.

Double-line tying method: Two lines are used. One line is connected to points A and B on the upper and lower parts of central rod respectively. The specific positions of points A and B vary according to the type of kite. A thief-catching knot is used for the two points. A half-folded knot is tied at point D for another line, and is connected to point A and point B at point C through the sling and bowknot after any excess thread end is cut off. The sling is connected to the thread end of the reel.

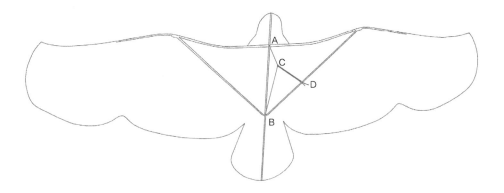

Three-line tying method: One line is connected to points A and B. The specific positions of the two points vary by kite. A thief-catching knot is used for the two points. The second line is connected to point E below the kite and point C in the middle of the first line. The specific position of point E varies by kite. A thief-catching knot is used for point E, and a bowknot is used for point C, so a knot is formed at point D, and a sling is formed between points C and D. A half-folded knot is tied at point G for the third line, and is connected to points D and E at point F through the sling and bowknot after any excess thread has been cut off. The sling is connected to the thread end of the reel.

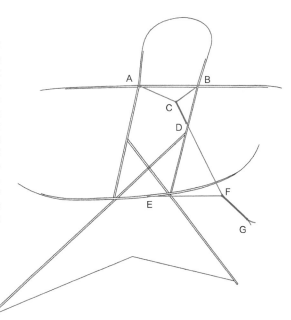

6. Understanding the framework structure diagram

It is very important to understand the framework structure diagram. The framework has two sides, and the framework structure diagram shows the line tying side, which is visible when the kite is flying. Let's take the eagle-shaped kite as an example.

The overall length, width and thickness of the kite framework are fixed. To change the size of a kite, you must adopt equal scaling. The framework structure diagram will be introduced in detail subsequently. After grasping the structure diagram, you will be able to make distinctive Chinese kites.

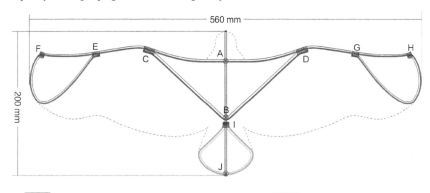

The lines of different colors enable you to understand the composition and shape of the kite framework. The specific specification of each component is marked behind its name:

■ Central rod   ■ Main wing strip   ■ Inclined prop   ■ Wing ring   ■ Tail strip

There are three different line types, each of which has a different meaning:

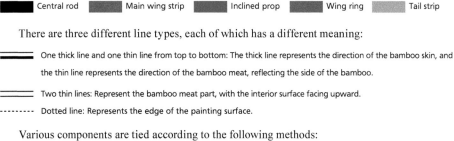

═══ One thick line and one thin line from top to bottom: The thick line represents the direction of the bamboo skin, and the thin line represents the direction of the bamboo meat, reflecting the side of the bamboo.

═══ Two thin lines: Represent the bamboo meat part, with the interior surface facing upward.

- - - - - - Dotted line: Represents the edge of the painting surface.

Various components are tied according to the following methods:

||||||||||| Densely parallel lines (parallel binding).

⊗ Cross patterns inside circle (Cross binding).

Kite line tying position:

● Red dot: Line positioning point.

Chapter 4 Basic Making Techniques   075

Chapter 5
Kite Cases

This chapter will expound on eight kinds of kites, including four board-like kites, three of them soft (one phoenix-shaped kite, one goldfish-shaped kite, one kite shaped like Monkey King's facial makeup) and one hard (eight trigrams-shaped kite), two soft-winged kites [one single-winged kite (eaglet-shaped kite) and one double-winged kite (butterfly-shaped kite)], two hard-winged kites [one rice-shaped hard-winged kite (with the theme of Hail the Judge) and one bank swallow-shaped kite (with the theme of happiness and longevity)].

Listed by making difficulty, board-like kites are followed by soft-winged kites and hard-winged kites. We will introduce their production methods in this order.

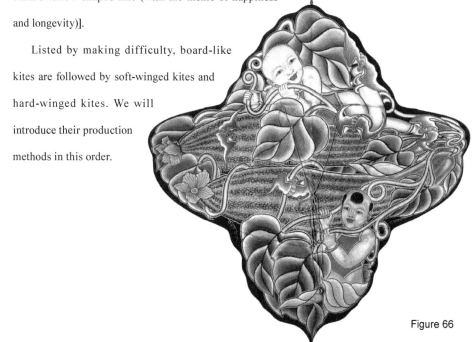

Figure 66

1. Phoenix-shaped kite

The phoenix is a holy bird and a totem of Chinese culture. According to legend, the phoenix reborn from fire has more powerful vitality, and is always regarded as a symbol of auspiciousness.

This is a typical soft board-like kite, whose four edges are made of paper and are movable. Its framework is like a simple cross. Two long-stripped tail belts made of painting surface material are pasted below the back of the painting surface, important for maintaining balance and the beauty of the kite flying.

Framework characteristics

The framework is very simple, consisting of a central rod and a main wing strip randomly stacked. Its biggest feature lies in that the two framework components are not connected through a line, but are pasted with glue, in consideration of very light weight of the kite.

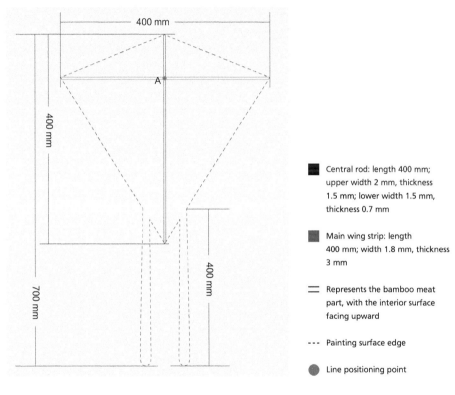

Central rod: length 400 mm; upper width 2 mm, thickness 1.5 mm; lower width 1.5 mm, thickness 0.7 mm

Main wing strip: length 400 mm; width 1.8 mm, thickness 3 mm

═ Represents the bamboo meat part, with the interior surface facing upward

--- Painting surface edge

● Line positioning point

Making the components and framework

Prepare two bamboo canes of the same length, and process the main wing strip with the all-in-one wing strip technique, and the central rod with the central rod technique. Connect the two bamboo canes in an overlapping way to form a cross shape, then paste and fix. Finish making of the framework.

Chapter 5 Kite Cases

Drawing the painting surface

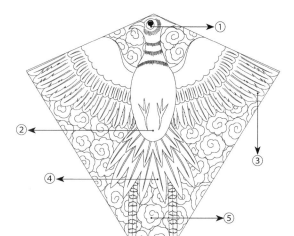

1. Draw the head of the phoenix with pigments in blue, red, yellow, and black.

2. Draw the body with pigments in red and black.

3. Draw the wings with pigments in blue, black, red, and white.

4. Draw the tail with pigments in green, red, and black.

5. Draw the background with pigments in blue and white, and make it look like a blue sky with white clouds.

6. Prepare another piece of white paper, and draw two ribbons attached to the phoenix tail with pigments in red and brown.

Pasting

1. Apply glue to the framework.

2. Attach the painting surface to the framework with applied glue, level the painting surface with your hands and cut off any excess white paper.

3. Cut off the excess white paper next to the tail belt.

4. Paste the tail belt under the kite according to the drawing.

5. Finish the pasting.

Tie lines

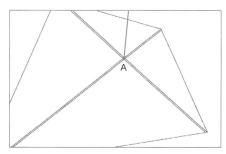

Use the single-line tying method to tie the line at point A.

Chapter 5 Kite Cases 81

2. Goldfish-shaped kite

Fish symbolizes good luck and happiness in Chinese culture. Fish (鱼) is homophonic to surplus (余), so almost every Chinese family eats fish or pastes paper-cuts or paintings of fish during the Spring Festival, praying for a rich and happy new year.

This is a typical soft board-like kite, whose four edges are made of paper and are movable. Its framework is shaped like a Chinese character "米". Its painting surface has the pattern of a very cute big-eyed green goldfish.

Framework characteristics

The kite framework consists of the central rod, main wing strip and inclined prop randomly stacked. Like the phoenix-shaped kite, it is also fixed with glue rather than line.

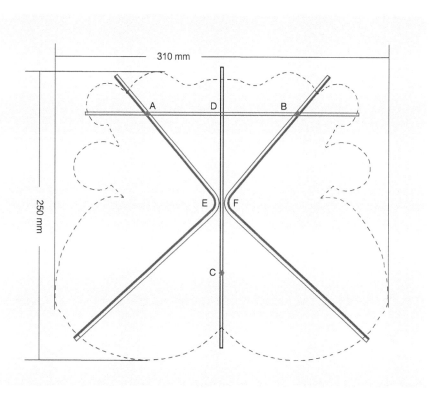

■ Central rod: length 245 mm; upper width 2 mm, thickness 1.5 mm; lower width 1.5 mm, thickness 0.7 mm

■ Main wing strip: length 255 mm; width 1.8 mm, thickness 3 mm

■ Inclined prop: length 340 mm; upper width 1 mm, thickness 2 mm; lower width 0.6 mm, thickness 1.5 mm

═ The thick line represents the direction of the bamboo skin, and the thin line represents the direction of the bamboo meat, reflecting the side of the bamboo.

═ Represents the bamboo meat part, with the interior surface facing upward

--- Painting surface edge

● Line positioning point

Making the components

Process the main wing strip with the all-in-one wing strip technique, the central rod with the central rod technique, and the inclined prop with the symmetrical splitting method, according to the size and shape marked in the drawing. Finish making the framework components.

Drawing the painting surface

1. Apply the fisheye and operculum with pigments in red, purple, yellow and black, and outline the thin edges with black pigment.

2. Apply the fish lip and nostril with pigments in red and light yellow.

3. Paint the pattern of Chinese copper coin in yellow and coffee between the two fisheyes.

4. Apply the fins in the thoracic and abdominal parts with pigments in green and yellow, and draw the outline with black pigment.

5. Apply pigments in green and yellow to the fish body, draw and outline the scales in black. Paint the dorsal fins in light green and dark green.

6. Apply the tail fins with green and yellow pigments and draw the outline in black.

Making the framework and pasting

1. Vertically connect the main wing strip and central rod in an overlapped way, paste and fix. Then paste the cruciform framework to the painting surface.

2. Connect the inclined prop to the main wing strip in an overlapping way, and make the inclined prop longer than the edge of the cloth cover. Bond the inclined prop and the painting surface without need of line tying.

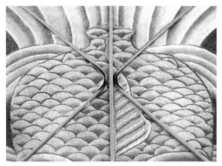

3. Make the inclined prop and central rod close to each other, but not connected.

4. Trim edges after pasting.

Tying the lines

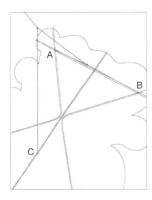

Use the three-line tying method to tie the line at points A, B, and C.

86 Chinese Kites

3. Kite shaped like the Monkey King's facial makeup

The Monkey King is the protagonist of *A Pilgrimage to the West*, one of China's four classical masterpieces. Having supernatural power and countless changes of tactics, he helps his master, the Tang Monk, to travel to the West to obtain Buddhist sutras. His legend is also depicted in Chinese Peking Opera. The facial makeup in Peking Opera is very particular, and the combination of different colors and patterns makes characters personalized. There is an inverted red peach pattern on the facial makeup of the Monkey King, whose outside is painted silver, which symbolizes his being loyal, brave and chivalrous.

This is a typical soft board-like kite, whose framework is like an umbrella. The pattern on the colorful painting surface portrays the agile and alert image of Monkey King. Likewise, two long-stripped tail belts made of painting surface material are pasted beside the back of the painting surface, to ensure balanced flying of the kite.

Framework characteristics

The kite framework shaped like an umbrella, consists of one central rod, one main wing strip, and two inclined props randomly stacked.

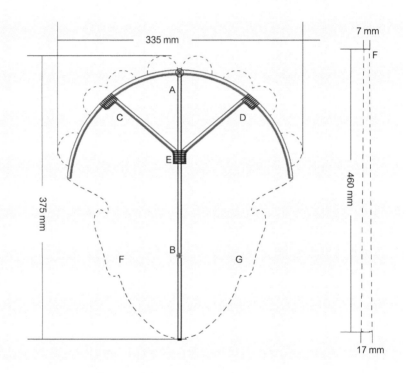

■ Central rod: length 350 mm; upper width 2 mm, thickness 1.5 mm; lower width 1.5 mm, thickness 0.7 mm

■ Inclined prop: length 140 mm; upper width 1 mm, thickness 2 mm; lower width 0.6 mm, thickness 1.5 mm

= Represent the bamboo meat part, with the interior surface facing upward

⊗ Cross binding

▨ Main wing strip: length 460 mm; middle thickness 3 mm, side thickness 2 mm, width 1.8 mm

= The thick line represents the direction of the bamboo skin, and the thin line represents the direction of the bamboo meat, reflecting the side of the bamboo.

--- Painting surface edge

||||| Parallel binding

● Line positioning point

Making the components

Process the main wing strip with the all-in-one wing strip technique, the central rod with the central rod technique, and the inclined prop with the symmetrical splitting method, according to the size and shape marked in the drawing. Finish making the framework components.

Making the framework

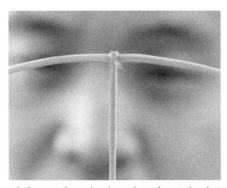

1. Connect the main wing strip and central rod at point A in a plug way, split the upper end of the central rod slightly, vertically clamp the main wing strip with the bamboo cane opening, and perform cross-binding to fix.

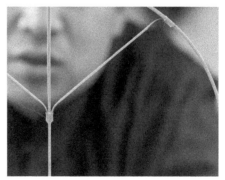

2. Use the kneeling connection method to connect the inclined props on both sides at points C, D and E of the main wing strip and central rod, and perform parallel binding to fix.

Drawing the painting surface

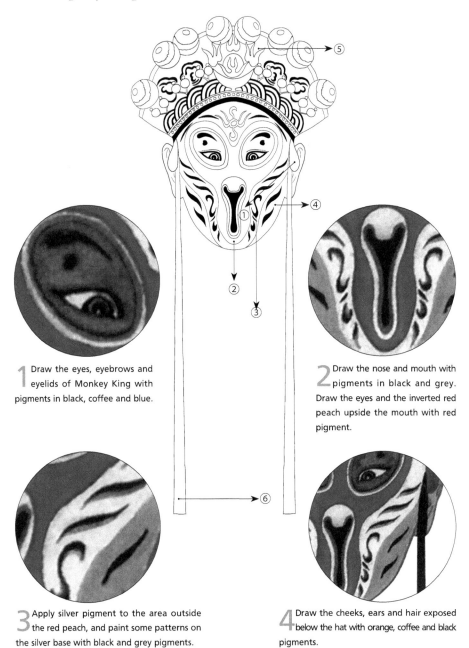

1. Draw the eyes, eyebrows and eyelids of Monkey King with pigments in black, coffee and blue.

2. Draw the nose and mouth with pigments in black and grey. Draw the eyes and the inverted red peach upside the mouth with red pigment.

3. Apply silver pigment to the area outside the red peach, and paint some patterns on the silver base with black and grey pigments.

4. Draw the cheeks, ears and hair exposed below the hat with orange, coffee and black pigments.

5. Draw the hat in yellow, red, blue, white, brown and black.

6. Draw the ribbon on the hat with black pigment.

Pasting

1. Apply glue to the framework, bond with the painting surface, and then trim the edges.

2. Paste the tail belt under the ear beside the back of the painting surface.

3. Finish.

Tie lines

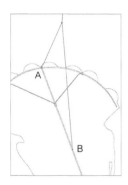

Use the two-line tying method to tie the lines at points A and B.

Chapter 5 Kite Cases

4. Eight trigrams-shaped kite

"Trigram" means the yin and yang representing the change of things. In the middle of the eight trigrams pattern is the fish representing the universe. The white fish denotes yang and the black fish indicates yin. The combination of white fish and black fish reveals the unity of yin and yang. The "—" and "--" outside the Eight Trigrams pattern represent yin and yang respectively, and three symbols in each group constitute eight different forms, namely the eight trigrams.

This is a typical hard board-like kite. Unlike soft board-like kites, its four edges are hard. It is also decorated with a string of colorful "Y"-shaped paper spikes, which are attached below the kite to ensure smooth flight.

Framework characteristics

The kite framework consists of one central rod, one central horizontal strip, eight edge strips, and 10 straight bamboo canes, which form two squares and a cross. Each square frame should be on the same level.

The 15 paper spikes below the kite are connected by two serial lines, which form a Y shape. The individual paper spike is 100 mm high and the space between paper spikes is 100 mm.

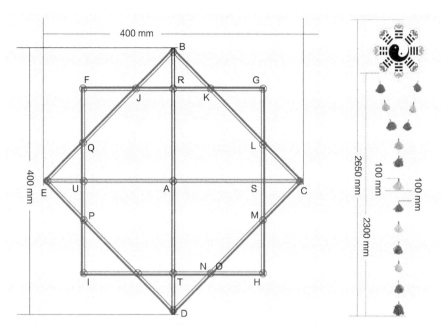

- ■ Central rod: length 400 mm; upper width 2 mm, thickness 1.5 mm; lower width 1.5 mm, thickness 0.7 mm
- ■ Central horizontal strip: length 400 mm; width 1.8 mm, middle thickness 3 mm
- ■ Edge strip: length 283 mm; upper width 1 mm, thickness 2 mm; lower width 0.6 mm, thickness 1.5 mm
- --- Painting surface edge

- ═ The thick line represents the direction of the bamboo skin, and the thin line represents the direction of the bamboo meat, reflecting the side of the bamboo.
- ═ Represents the bamboo meat part, with the interior surface facing upward
- ⊗ Cross binding
- ● Line positioning point

Making the components

Make 10 linear bamboo components according to the size and shape marked in the drawing. Eight edge strips must be consistent in width, thickness and hardness. Certain excess length can be retained for easy bonding and line tying.

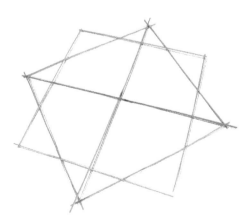

Chapter 5 Kite Cases 93

Making the framework

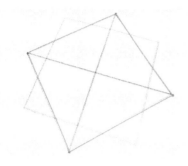

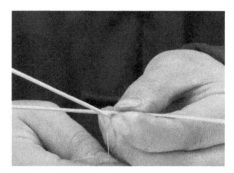

1. Split the ends of the four edge strips separately, connect them into a square through the butt clamp connection method, and perform cross binding after pasting.

2. Repeated step 1 again to make a square. As shown in the drawing, connect the eight intersection points (J, K, L, M, N, O, P, Q) of the two squares in an overlapping way, and perform cross binding to fix.

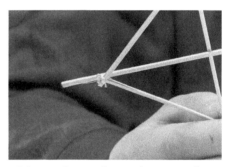

3. Connect the central rod and the central horizontal strip in an overlapping way, and then perform cross binding to form a cross shape.

4. Connect the cross shape and two square frames in an overlapping way, and perform cross binding to fix.

5. Fix all the connection points with 502 glue. After the glue solidifies, use the side-cutting plier to cut off the excess bamboo canes.

6. Finish making the framework.

Draw the painting surface

1. Draw a yin-yang symbol on the center of the painting surface with black ink.

2. Draw eight diagrams around the yin-yang symbol.

3. Draw a green bat on each of the four corners of a square.

4. Draw a red bat on each of the four corners of the other square.

Chapter 5 Kite Cases

Pasting

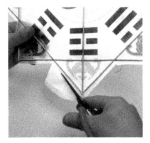  

1. Apply glue to the framework and then paste the painting surface on the framework. Cut off the excess part of the painting surface and reserve enough width for edge covering.

2. The framework of an eight trigrams-shaped kite is linear, and can be directly glued and edge covered.

3. Finish pasting.

Making the long spike

1. Take out a piece of red crepe paper 10 cm wide and about 20 cm long, fold it into four layers, and evenly cut two-thirds of the lower part into strips.

2. Wind up after inserting the serial line into the paper spike, and tighten the head of the paper spike with another line.

3. Take out a piece of yellow crepe paper about 3 cm wide and about 5 cm long, wrap the curled head part of red paper spike, and apply glue to fix instead of tying lines.

4. Make a total of 15 paper spikes in the same way, and arrange different colors of paper spikes at intervals.

96 Chinese Kites

Tying the lines

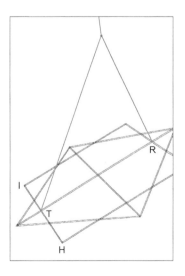

Use the double-line tying method to tie the lines at points R and T. Dig a small hole respectively at points I and H on the inner painting surface, and pass the rope for the paper spike string through the holes, and tie to the bamboo canes. Finish making the kite.

Chapter 5 Kite Cases

References

1. Wang Hongxun and Xu Shuyan. *Chinese Kites*, Heilongjiang Fine Arts Publishing House, 1994.
2. Fei Baoling. *Study of Cao Xueqin's Paintings about Swallow-Shaped Kites*, ECHO of Things Chinese, 1998.
3. Wang Qinian, Wu Guanghui and Yu Jiming. *Kites*, People's Sports Publishing House of China, 1986.
4. Kong Xiangze, Kong Lingmin and Kong Bingzhang. *Cao Xueqin's Kite Art*, Beijing Industrial Art Publishing House, 2004.
5. Ha Kuiming and Ha Yiqi. *Ha Kites of China*, Commercial Press Hong Kong Branch, 1986.
6. Feng Jicai. *Folk Kites*, Hebei Children's Publishing House, 2004.

图书在版编目（CIP）数据

　　风筝：英文 / 刘宾著；周灵译. -- 北京：中国画报出版社, 2019.12
　　（手工中国风）
　　ISBN 978-7-5146-1791-7

　　Ⅰ. ①风… Ⅱ. ①刘… ②周… Ⅲ. ①风筝—制作—中国—英文 Ⅳ. ①TS938.91

中国版本图书馆CIP数据核字(2019)第179208号

风筝（英文）

刘宾　著

出 版 人：于九涛
策划编辑：刘晓雪
责任编辑：刘晓雪
英文翻译：周　灵
英文编辑：王国振　陈　旭
英文改稿：迈克尔·杰佛里·默里（Michael Geoffrey Murray）
英文校对：朱露茜
美术设计：赵艳超
责任印制：焦　洋

出版发行：中国画报出版社
地　　址：中国北京市海淀区车公庄西路33号　邮编：100048
发 行 部：010-68469781　010-68414683（传真）
总 编 室：010-88417359　版权部：010-88417359

开　　本：16开（710mm×1000mm）
印　　张：7
字　　数：70千字
版　　次：2019年12月第1版　2019年12月第1次印刷
印　　刷：北京京华虎彩印刷有限公司
书　　号：ISBN 978-7-5146-1791-7
定　　价：68.00元